Emil Rosenberg

# The use of the spectroscope in its application to scientific and practical medicine

Emil Rosenberg

**The use of the spectroscope in its application to scientific and practical medicine**

ISBN/EAN: 9783742821621

Manufactured in Europe, USA, Canada, Australia, Japa

Cover: Foto ©Lupo / pixelio.de

Manufactured and distributed by brebook publishing software
(www.brebook.com)

Emil Rosenberg

# The use of the spectroscope in its application to scientific and practical medicine

# CONTENTS.

# DESCRIPTION OF THE TABLE.

1. Solar Spectrum.
2. Chloride of Cobalt solution (in absolute alcohol).
3. Blue glass.
4. Chrom-alum (sol. in water).
5. Ammonio-sulphate of copper (in water).
6. Prussian blue (in oxalic acid).
7. Indigo (in sulph. acid).
8. Chloride of copper (in water).
9. Permanganate of potassium.
10. Aniline-Red.
11. Bichrom. of potassium (in water).
12. Green glass.
13. Bergamot-oil.
14. Chlorophyll (in ether).
15. Tinctura Aconiti ex foliis.
16. Carmine (ammoniacal solution).
17. Oxyhæmoglobin.
18. Hæmoglobin (Reduction-band).
19. Hæmatin (Oxyhæmatin).
20. Hæmin (Acid-band).
21. Reduced Hæmatin.
22. Methæmoglobin.
23. Blood treated by SNa in excess.
24. Fel tauri inspissatum (in muriatic acid).
25. Urobilin, acid solution.
26. Urobilin, alkaline solution.

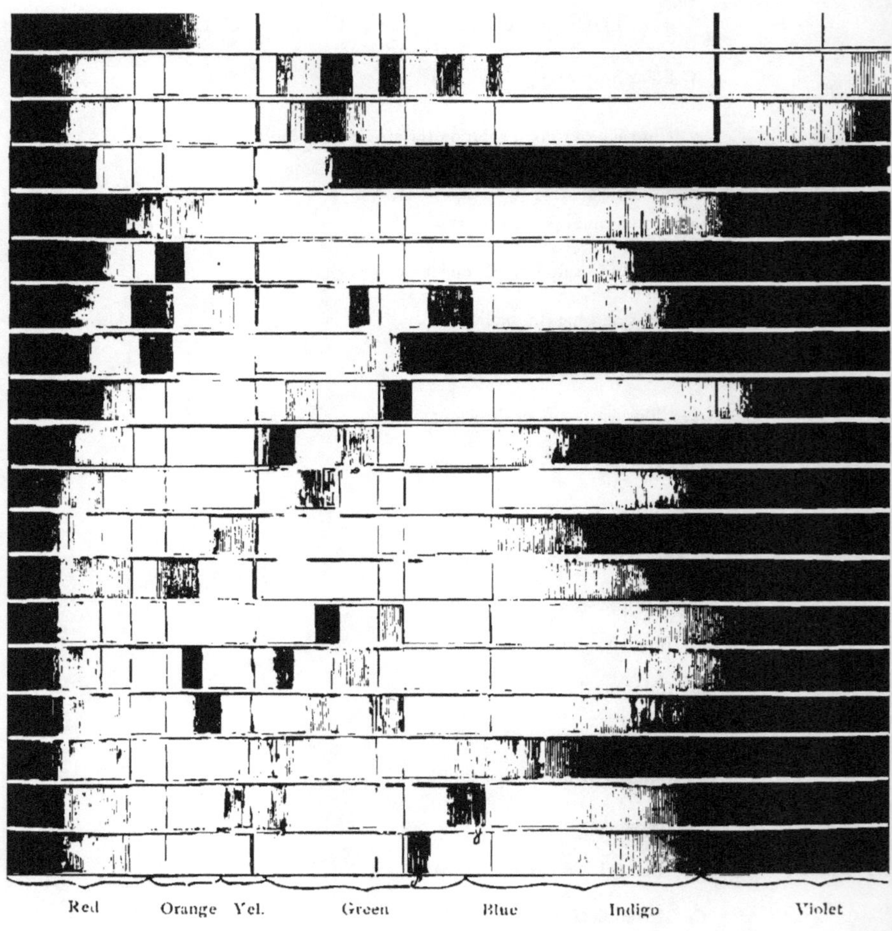
Red
Orange
Yel.
Green
Blue
Indigo
Violet

# THE USE OF THE SPECTROSCOPE.

## CHAPTER I.

### THE APPARATUS AND METHOD OF USE.

IT is not proposed, in this brief essay, to give a complete exposition of all the optical relations of our subject, as this want has been thoroughly met in the masterly productions of Schellen, Roscoe, and Lockyer, who have fully and satisfactorily established themselves as the authorized and acknowledged exponents of the general principles of Spectrum Analysis. But although in these works the apparatus, with all its varieties, is fully described, we think it necessary to reproduce here a detailed description of the instrument and the mode of using it.

The principal part of the Spectroscope (that by which the decomposition of light is effected) is either a grating or a prism. We shall confine ourselves to the discussion of the prism, which has led to such extensive discoveries; though it must be conceded that it is only by means of the diffraction-grating that a truly normal spectrum can be obtained.

The prisms used are either solid or hollow, the latter being filled with a fluid of high refractive power (bisulphide of carbon, $CS_2$). One or more prisms may be employed at the same time. The single prism may be either simple or compound. The simple glass (dense flint, of a refracting angle of 60°), or the ordinary $CS_2$ prism, effect a deflection (deviation) and dispersion of the incident ray, allowing only oblique vision. The compound glass prism (a combination of several flint and crown glass prisms) enables the refracted ray to emerge in the straight continuation of the incident beam. These compound prisms, therefore, permit direct vision, the spectrum being seen by looking straight at the source of light. During the last two years we have been in possession of a compound $CS_2$ prism. The inventor, Prof. A. K. Eaton, of Brooklyn, N. Y., conceived the idea of combining the $CS_2$ prism with a piece of crown glass, which combination, of course, also permits direct vision.

Although a prism is, in a certain sense, a Spectroscope *per se*, there are, besides, in the construction of a complete apparatus, several other essential parts:

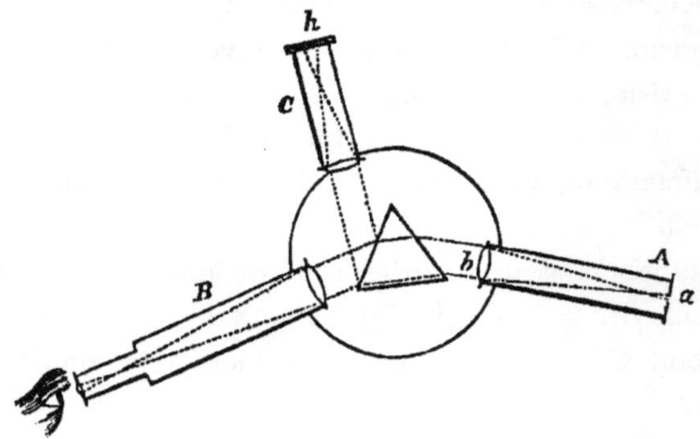


FIG. 1.

1. A vertical slit, the breadth of which can be regulated by means of a screw, at the end (*a*) of the collimating tube (*A*); at the other end, nearest the prism, an achromatic lens (the Collimator) is placed at the distance of its focus for rays from the slit, which are made parallel by the Collimator and thrown upon the prism.

2. An astronomical telescope (*B*) of low magnifying power, through which the rays of light, refracted and decomposed into a spectrum, reach the eye of the observer.
3. For fixing the position of spectral lines or bands, various auxiliary contrivances are employed; thus, for instance, Kirchhoff and Bunsen have introduced, at the head of a third tube (*C*), a reduced photographed millimetre scale (*h*), which can be seen through the telescope by reflection from the front surface of the prism. The three tubes are fastened to a central plate, carrying the prism, and screwed to the upper end of a cast-iron foot. The Collimator-tube (or the arm supporting it) must be immovable. The telescope and scale-tubes are so fitted that they are movable in a horizontal plane about the axis of the base. In front of the slit is a small equilateral glass prism, covering the lower half of the slit. This prism of comparison transmits by total reflection the light of a second lateral source through the slit, so that the observer sees the spectra of the two sources of light, one under the other.

The arrangement of such an (Bunsen and Kirchhoff) apparatus is effected in the following manner:

The prism is taken off the stand, and the observer looks first through the collimator-tube in the direction *b* to *a*, with moderately opened slit, the slit-tube being drawn out, until the edges of the slit appear sharply defined. Then, the telescope, having been adjusted for a distant object, is brought into position in a straight line with the collimator-tube, and the slit observed through the telescope; the latter has to be moved, so that the slit will appear as a sharp bright line in the middle of the field of view. The prism is now brought again to its place. The scale-tube is turned through an angle round the axis of the base, and the illuminated scale is pushed in with its tube, until the division of the scale, observed through the telescope, is distinctly perceived. Stray light is excluded by covering the prism and the ends of the tubes adjoining it with a black cloth.

The illumination of the scale is effected by means of a light placed before it. A white paper screen interposed between the flame and the scale will obviate the unpleasant glare; and often the reflex of the diffused light of the room from a white paper will be found sufficient for the illumination of the scale without an additional light.

Other apparatus* differ from the instrument just described in being provided with only two tubes; the scale being placed in the telescope, just behind the ocular.

The central plate may easily be arranged for prisms of different kinds. When a simple glass or $CS_2$ prism is employed, the telescope will have to be passed through a certain angle round the axis of the base. The compound glass, or Eaton's compound $CS_2$ prism will, on the contrary, require both tubes (the telescope and the collimating-tube) to be in a nearly straight line. In this case we should have a direct-vision Spectroscope.

The prism has to be placed so that the edge of the refracting angle is parallel to the slit. Besides, the position of the prism depends upon the degree of the dispersion required; and, in order to obtain a pure spectrum, it must be that of "the angle of minimum deviation."

For the fuller explanation of the latter point, we refer to the subjoined diagram,

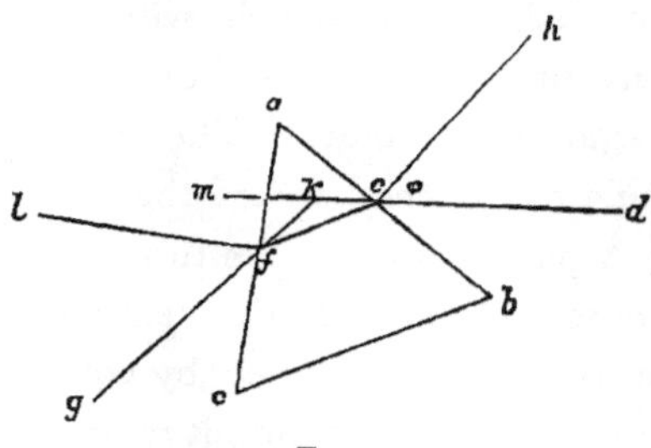


FIG. 2.

in which *a b c* represents a prism, *a b* and *a c* its refracting faces, *d e* the incident ray, which is refracted within the prism as *e f*, and then again emerging as *f g*. The recip-

* The instruments made here, in New York, have a mirror attached to the Collimator, with two movements for reflecting the rays of the sun into the slit.

rocal direction of the incident and the emerging ray depends (other conditions being equal) upon the angle of incidence $\phi$ (formed by *d e* with *h e*, the perpendicular upon *a b*).

This angle varies with the position of the prism. By turning the prism around the axis of its length, a position will be obtained, in which the angle of incidence will be equal to the angle of refraction ($d\ e\ h = g\ f\ l$), and in which the angle (*m k g*) formed by the prolongation of the emerging and of the original path of the incident ray within the prism, acquires its minimum value. The angle *m k g*, measuring the amount of deviation of the prismatic image, shows the deflection to be reduced to a minimum, and is therefore called the angle of minimum deviation.

Practically, the position of minimum deviation is found by slowly turning the prism round its axis, until the Fraunhofer's lines or the bright metal lines will exhibit the least perceptible change as to their location. Besides it is the precision * with which certain definite lines are perceived—for instance, the Fraunhofer's lines *a*, *B*, *C*, that indicates whether the minimal deviation is in accordance with their respective region—in this instance with the Red. From all this it would follow, that the simple prism has to be brought into various positions, according to each separate line which we wish to observe. But in practice this will be found to be unnecessary.

In addition to the ordinary numbered scale, there are other methods in use for the localization of spectrum lines or bands.

A well defined mark of some kind, such as a wire or

* In addition to all the above-described procedures, it is usually necessary to turn the whole instrument slightly around the axis of its length, in order to obtain Fraunhofer's or metal lines with greatest precision.

cross-wires, or a line of light, are made to traverse the spectrum; and the amount of motion is measured externally by means of a micrometrical arrangement; or the position of that part of the spectrum covered by the cross-wires may be read off a horizontal graduated circle provided for that purpose.

Another method for fixing the location of appearances in the spectrum has more recently been devised by Professor J. C. Dalton.* Its chief feature is a photographed scale, presenting not only divisions, but the position of the Fraunhofer's lines permanently marked—of course for a certain permanently fixed position of the several parts of the instrument.

Among the varieties of Spectrum Apparatus, there is one of great importance for the physiologist and physician; viz. the micro-spectroscope, by which the absorption-phenomena of the most minute solid and liquid bodies, for instance, that of a corpuscle of blood, may be accurately observed. The spectroscopical part of such an instrument (Sorby-Browning) can be applied to any microscope by fixing it in the place of the ordinary eye-piece. A complete description of the micro-spectroscope may be found in the works of Schellen or Roscoe.

Before commencing practical operations with the Spectroscope, it is necessary to determine—in relation to the scale employed—the position of the principal Fraunhofer's lines † (*A* to *H*), as they appear in the sun spectrum, if a narrow slit is used. (With a widely opened slit, the sun will

* Dalton, J. C., A new method of determining the position of absorption-bands in the spectrum of colored organic fluids. (A paper read before the N. Y. Academy of Medicine, in October, 1874.)

† See note at the end of this chapter.

produce a continuous spectrum, as Newton saw it). A description of the scale thus formed must precede every report of spectral observations, for the reason that each spectrum, obtained by prismatic refraction, is always only an individual one. Such a "scale" may be chosen once for all and retained.

The simplest method will be to determine all the principal Fraunhofer's lines by direct observation in sunlight; the position of the prism and its distance from the collimating lens being fixed. Both may be easily marked on the instrument, or made permanent by special contrivances. Even when there is but little sunlight, the lines D, E, *b*, and F are determinable by direct observation; the others are readily calculated by Kirchhoff's or any other exact scale.

The Fraunhofer's lines may, on the other hand, be determined independently of the sun. Thus Heynsius and Campbell * have adopted the following rather complicated method. In the first place D is fixed by means of the Sodium line Na, coincident with D; and A through the dark red Potassium line K$\alpha$, coincident with A. Then the position of C and F is found by means of a Geissler tube filled with hydrogen gas.† After that, the lines a, B, E, and *b* have to be calculated according to "Schellen's scale."

Occasionally, the scale is made only with reference to some metal lines. Thus the scale chosen may be sufficiently defined by mentioning the position of the lines K$\alpha$, Li $\alpha$, Na, Sr $\delta$.

The employment of metal lines is absolutely indispensable in controlling and verifying the position chosen in day-

* Archiv f. Physiologie, iv. 1871, p. 520.

† Electric sparks passing through such a tube, containing hydrogen at the pressure of one atmosphere, produce a bright carmine red of the incandescent hydrogen, which color is resolved by the spectrum into a red line (coincident with C), a bluish-green line (coincident with F), and a blue line (coincident with G).

light, when artificial light is used. The bright yellow Sodium line is especially employed for this purpose.

The scale presented below is obtained by a simple flint prism of 60°, used by the author; the dispersion of which in a certain position being so small as to allow the entire spectrum to be brought into the compass of the scale:

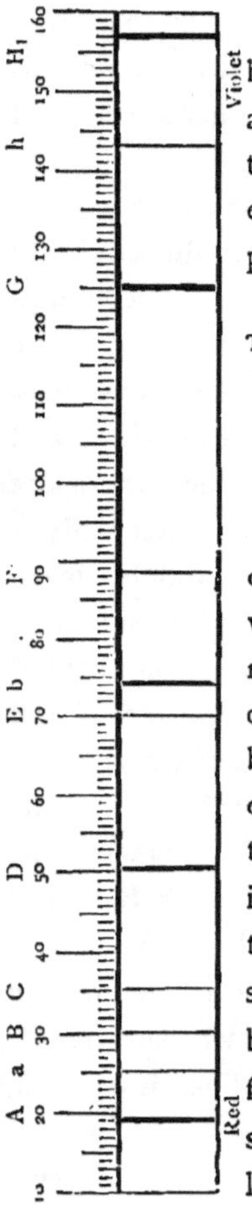

Fig. 3.

The red end of the spectrum is brought to the observer's left hand, by placing the refracting edge of the prism to the left. The spectrum observed through the prism without the telescope, presents in this position the red end to the right—through inversion by the telescope it is seen at the left.

When a greater dispersion is required, or the spectrum is displaced towards the violet end, in order to bring the red end nearer to the centre of the field of view, only a portion of the spectrum will be visible at one time; the remaining sections can only be brought into view by moving the telescope horizontally. When artificial light is used, one of the metal lines occurring in the blue section of the spectrum, for instance, the blue Strontium line, Sr $\delta$, or the blue Caesium line, Cs $a$, can be employed for fixing F and G, provided the distance of such a metal line from both Fraunhofer's lines has been determined in daylight, and the relative position of the several parts of the instrument retained. If the dispersion is not too great, the Thallium line, or even the Na line may be employed for the same purpose.

(In the three-tube instrument, the Collimator-tube, the prism and the scale-tube are the three parts, the reciprocal position of which has to remain unaltered, if the once-chosen scale is to be retained.)

In reference to the calculation previously alluded to, we insert in this place Kirchhoff and Bunsen's Reduced and Larger scales:

| KIRCHHOFF'S LARGE SCALE. | | KIRCHHOFF'S REDUCED SCALE. | |
|---|---|---|---|
| A | = 405 | A | = 17.5 |
| a | = 505 | a | = 23 |
| B | = 593 | B | = 28 |
| C | = 694 | C | = 34 |
| $D_1$ } | = 1002.8 | D | = 50 |
| $D_2$ } | = 1006.8 | | |
| E | = 1522.7 | E | = 71 |
| $b_1$ } | = 1633.4 | b | = 75.9 |
| $b_2$ } | = 1648.3 | | |
| $b_3$ } | = 1655.0 | | |
| F | = 2080 | F | = 89.8 |
| G | = 2855 | G | = 127.5 |
| *h* | = 3371 | *h* | = 148 |
| H | = 3568 | $H_1$ | = 161.9 |
| | | $H_2$ | = 166 |

Preyer's scale ("Die Blutkrystalle," p. 227,) may also be profitably employed for this purpose:

PREYER'S SCALE.

A = 31.2 D = 59.5
a = 36.5 E = 77.9
B = 41.0 b = 81.3 ($b_1$ = 81.0, and $b_4$ = 81.6)
C = 45.9 F = 94.6
G = 128.5 (128.0—129.0)
$H_1$ = 158.7 (158.2—159.2)

*** The measures refer to the centre of the lines or groups.

Though the calculation is very simple, it may not be out of place to illustrate it by an example:

Suppose D, E, *b* and F having been observed by sunlight, A would be found by this rule:

The distance DE (our scale) : $x$ (the distance AD) = the distance DE (Kirchhoff) : the distance AD (Kirchhoff.)

In order to attain the necessary exactness in using the bright lines as substitutes for Fraunhofer's lines, the following point has to be considered. The bright spectrum lines will always show the width of the slit. In widening the slit by means of the screw only one edge is displaced, the other being immovable. If the latter be to the right hand of the prism, the left edge of the bright line will—through the inversion by the telescope—remain at its place, while the expansion of the bright band will take place towards the right.

We would therefore have, in this case, to bring the *left* edge (for instance, of the yellow Sodium line) under the number of the scale chosen for the line D.

Besides the vertical Fraunhofer's lines, the observer will frequently see horizontal lines. These are produced by inequalities of the slit or by dust. By cleansing the slit with a small brush or piece of paper they will necessarily in the latter instance disappear. They are, however, useful in adjusting the distance of the slit from the prism, which is only correct, if the vertical Fraunhofer's lines, the horizontal dust lines, and the divisions of the projected scale (or the cross-wires) are perceived with equal distinctness. It is evident that these dust lines will be even more useful in this respect in the absence of Fraunhofer's lines, viz., when artificial light is employed.

## THE FRAUNHOFER'S LINES.

Each glowing gas * absorbs exclusively rays of the same refrangibility as those which it would itself emit (radiate) in the state of incandescence.

The Emission-(radiation-) spectrum of each gas glowing in the solar atmosphere, for instance of hydrogen, which consists of the three bright lines already mentioned, must be reversed (appear black) if rays are transmitted through it from a source of light sufficiently bright, and giving for itself a continuous spectrum (the white-hot central body of the sun).

* Metal molecules glowing in flames or in the electric spark, and producing thereby the bright metal lines, are also in gas form.

# CHAPTER II.

## EMISSION-SPECTRA, AND THEIR PRACTICAL APPLICATION TO MEDICINE.

SOLID and fluid substances emit in the state of incandescence, light of *every* degree of refrangibility; that is, they give a continuous spectrum.* An exception to this law is presented by solid Erbia, which exhibits bright lines.

Each body emits in the condition of glowing gas, light of *various* degrees of refrangibility, giving a discontinuous spectrum which contain bright bands or lines upon a dark background.

The incandescent vapors of the alkalies and alkaline earths: Sodium, Potassium, Lithium, Calcium, Strontium, Barium, Rubidium, Caesium, Thallium, and Indium, give discontinuous spectra, for the most part visible by means of the spirit lamp, or through the non-luminous flame of the Bunsen burner (or the hydrogen flame). The heavy metals (metals proper) require the heat of the electric spark for their volatilization.

We have seen that a knowledge of some of the bright metal lines is necessary, if only for the determination of the Fraunhofer's lines when artificial light is used.

* Law discovered by J. W. Draper: "In view of the foregoing facts, I conclude that, as the temperature of an incandescent body rises, it emits rays of light of an increasing refrangibility." (On the production of Light by Heat. Phil. Mag. xxx. 1847, p. 345.)

In medicine, the Emission-Spectrum Analysis is applied to the establishment, in animal (or vegetable) tissues, of the normal or accidental occurrence of chemical elements, which might not be demonstrable by ordinary chemical analysis with the same facility, or the identification of which might be impossible, from the exceedingly small quantity in which they may be present.

For the purpose of examination, the animal tissues are brought into the flame, either fresh, so as to be carbonized and partly incinerated, or they may first be reduced to ashes (in the platinum crucible) and thus examined without further preparation. Another method consists in treating the coal or ashes by pure muriatic acid, so as to transform the alkalies into their chlorides, which are volatilized with greater facility. The details of the procedure are as follows:

The end of a thin wire of platinum is looped, and heated until it ceases to give a luminous flame; a bead is then melted into the loop by taking up a small quantity of the ashes with the moistened wire, and drying it at the surface of the flame. The Bunsen burner is then placed before the slit at a distance of about 5 to 10 centim., the upper edge of the chimney being 2 or 3 centim. beneath the lower end of the slit. The air is shut off below, so as to permit a brilliant spectrum to appear through the telescope, which determines the right position for the burner. The air is then passed through the openings into the mixing-chamber, the scale is illuminated, and the platinum wire, fastened to a glass tube, is brought into the outer mantle of the non-luminous flame.

In observing the spectra of alkaline chlorides in general (from solutions), the contrivance known as "Mitscherlich's tube or wick" will be found much more convenient than the thin wire.*

We propose now to give a description of the characteristic features of these spectra; referring, for more specific details, to the abstract and drawings, after Bunsen and Kirchhoff, given in Roscoe's work.†

Sodium (Na) occurs in the animal economy, more especially in bloodplasma, urine, pancreatic fluid, the bile of man and the majority of animals, and also in serous transudations. It imparts to the non-luminous flame a bright yellow color, and presents a spectrum consisting of only one yellow line, which is coincident with Fraunhofer's dark line D. (These notes refer only to the spectra of the chlorides, as obtained by means of the Bunsen burner.)

Potassium (K) is more especially present in the ashes of muscle, of red blood corpuscles, nerves, yolk of egg, and milk. It gives a pale violet color to flame, and the Spectrum presents two characteristic lines:

K*α*‡, dark red, coincident with the dark line A of the solar spectrum. §

K*β*, violet, very near H.

The Potassium lines are often concealed by the intensity of the yellow line of Sodium, which mostly appears

* The author finds ordinary steel-pens a good substitute for the wire. The point being fixed in the holder, the pen forms a convenient metal-spoon for subjecting traces of the chlorides to the heat of the Bunsen burner.

† Kirchhoff and Bunsen, "Chemical Analysis by Spectrum Observations." First Memoir: Pogg. Ann. CX. 161. Phil. Mag., 4th series (1860) xx. 89. Second Memoir: Phil. Mag. xxii. 1861. Abstract in Roscoe, 1869, p. 68.

‡ The Greek letters designate the intensity, *α* signifying the brightest line.

§ M. Louis Grandeau (Instruction pratique sur l'Analyse Spectrale, Paris, 1863—a book which is to be highly recommended) declares on the contrary the line K*α* not to be coincident with any Fraunhofer's line. Either is correct. K*α* coincides with A, if the observation is made with a simple apparatus and with a telescope of low magnifying power. It appears less refrangible than A, if observed by the electric spark, several prisms and high magnifying power (Kirchhoff).

simultaneously. By interposing a cobalt solution (Table, Fig. 2), or a blue cobalt glass, the intense Sodium light may in such a case be weakened; by interposing an Indigo solution (Fig. 7,) it will be entirely obscured. The Potassium spectrum has the peculiarity of being continuous in the middle part, from yellow to blue.*

Lithium (Li) has been found by means of its spectrum in traces in muscle, blood, and milk of the human subject and animals. The flame is of a crimson color, and shows a spectrum consisting of a characteristic bright red line, Li $\alpha$, between B and C, and a second very faint yellow line, Li $\beta$, between C and D.

Calcium (Ca) is largely found in bone and teeth; in smaller quantity in any animal fluid, and many pathological products. The flame is of a yellowish-red color, and shows a spectrum consisting of nine lines, of which two are characteristic:

Ca $\alpha$ a broad orange line between C and D;

Ca $\beta$, a broad bright green line between D and E.

Strontium (Sr) is introduced into the animal system by food or water containing it. It colors the flame red. The spectrum consists of eight lines, four of which are characteristic:

Sr $\alpha$, Sr $\beta$, Sr $\gamma$, are red lines, between C and D, the first named close by D, or the Sodium line.

Sr $\delta$ is blue, between F and G.

Barium (Ba) exceptionally occurs in the body, from the same sources as Strontium; it produces in flame a

* The Potassium lines are very easily and conveniently perceived, by holding a grape-stem (tartrate of potassa) before the slit in the non-luminous flame of the Bunsen burner.

Matches, prepared with chlorate of potash, will likewise be found very useful in this respect.

green tint, and gives a spectrum consisting of fifteen lines, three of which are characteristic:

Ba $\alpha$, a broad green band between E and *b*.

Ba $\beta$, near *b*, towards the blue; and

Ba $\gamma$, between D and E.

Thallium (Tl) colors the flame green. The spectrum consists of one bright green line between D and E, nearer E.

Indium (In) shows a violet-colored flame. The spectrum consists of two lines:

In $\alpha$, between F and G, blue.

In $\beta$, between G and H, violet.

Rubidium (Rb) exhibits a reddish-blue flame. The spectrum comprises ten lines, four of which are characteristic:

Rb $\alpha$, and Rb $\beta$ are violet lines, between G and H.

Rb $\gamma$, and Rb $\delta$ are red lines, which fall beyond Fraunhofer's line A.

Caesium * (Cs) presents a similar-colored flame, the spectrum embracing eleven lines, three of which are characteristic:

Cs $\alpha$ and Cs $\beta$, blue lines between F and G.

Cs $\gamma$, red, between C and D.

There is no difficulty in producing a compound spectrum of K, Na, Li and Ca from the tissues of man or mammals.

Every tissue which is rich in blood-vessels, will yield the lines of one of these metals, even if the latter be not contained in the tissue itself, but only in the blood.

It is therefore obvious, that bloodless tissues, destined for Spectrum Analysis, must be separated at a perceptible distance from their matrix.

Lithium has been found in parings of nails, and in the

* Rubidium and Caesium possess the same peculiarity as Potassium, giving a continuous spectrum in the centre.

hair of living man. Hence it occurs independently from any admixture of blood.*

An application has been made by Professor Valentin,† of one of these metal lines to physiological research. He employed the Strontium spectrum as a substitute for the chemical (Hering-Vierordt's) ferrocyanide of potassium test, for the purpose of measuring the velocity of absorption or of the circulation. Sodium, Potassium, etc., being normal constituents of the system, can not be so employed.

Strontium, however, being in exceptional cases introduced into the body, a previous test of its presence with a drop of the blood of the animal to be experimented upon is required. But even should its accidental presence in minute traces be proved, it is only necessary to use a weaker flame, which, under the circumstances, would not produce the spectrum.

(For an application of the bright metal lines in Ophthalmology, see Appendix.)

In reference to the demonstration of heavy metals, we find in Kirchhoff's tables ‡ the lines of Zinc (Zn), Nickel (Ni), Magnesium (Mg), Gold (Au), Tin (Sn), Mercury (Hg), Copper (Cu), Iron (Fe), Aluminium (Al), Arsenic (As), Stibium (Sb), Lead (Pb), Cobalt (Co), Silver (Ag), Chromium (Cr), and Cadmium (Cd). Among these are some of the most ordinary poisoning agents. But, although Valentin, as early as 1863, declared that "the application of this means to the diagnosis of metal poisoning will undoubtedly

* For spectral-analytical demonstration of Tl and Li in urine, see Neubauer und Vogel, Analyse des Harns, 6th edition, 1872, pp. 134 and 135.

† Valentin, G. Der Gebrauch des Spektroscopes zu physiologischen und ärztlichen Zwecken. Leipzig, 1863.

‡ Compare also Wm. Huggins' maps of the metallic lines. These, as well as Kirchhoff's and Angstrom's, are found in Roscoe.

form one of the brightest features in the medical use of the spectroscope,"—we must confess that there is, so far, not a single observation on record concerning the examination by the spectrum of animal tissues impregnated by such poisons, —not even an experimental one. The temperature of the Bunsen burner, notwithstanding its 2350° C., is not sufficient for the volatilization of the heavy metals (with the exception of copper and manganese). Their conversion into incandescent vapors must be effected by the electric spark, as produced by 20 to 40 Bunsen elements, charged with nitric acid, or with a weaker battery, by the aid of the Ruhmkorff Inductor, sometimes strengthened by the interpolation of a Leyden jar. The animal tissues should for this purpose be placed in a small, flat coal crucible, to which the spark from the other charcoal-point is made to pass over.

As to the Emission-Spectra of the gases and non-metallic elements (Air, Hydrogen, Nitrogen, Oxygen, Sulphur, Selenium, Phosphorus, Chlorine, Bromine, Iodine), we must confine ourselves to the mention of their production by electric discharge, through Geissler tubes filled with the rarefied gases. That these spectra are vastly important in spectroscopic study in general, has been elsewhere shown in reference to Hydrogen.

Regarding their application in medicine, Valentin's proposition may be of interest to the student of comparative physiology. He says*: "These gas-spectra, especially those of hydrogen and carbonic acid, might be employed for qualitative Analysis of gases, existing only in minute quantity in several parts of the animal economy—as the intestinal gases of the frog, insects, and other creatures. The gases would have, however, before introducing them

* Loc. cit., p. 131.

into a Geissler tube, to be dried thoroughly with sulphuric or phosphoric acid, and divested of Ammonia, if it be present; and, besides, the atmosphere would have to be rigidly excluded during the filling of the tube."

If Spectrum-Analysis has so far been successfully applied only to the discovery of the alkaline metals and earths, while the spectra of the metals proper, from their complicated nature and other difficulties, seem not destined to become of great practical value to the physiologist—it must be observed, that it is the former, that are with the greatest difficulty distinguished by ordinary chemical reactions, especially if—as is generally the case—only present in minute quantity in a mixture.

## CHAPTER III.

### ABSORPTION SPECTRA.—BLOOD.

NOTWITHSTANDING the important facts offered in medical Spectrum Analysis by the Emission phenomena, yet the main subject for our consideration is that section relating to spectral appearances of transparent colored bodies, more especially fluids.*

If sunlight, or the light of a bright flame be passed vertically through a colored solution, placed immediately before the slit in a glass vessel with plane-parallel walls, a great number of *pigments* will, by the gradual dilution of the fluid, present a discontinuous absorptive spectrum, certain definite sections thereof being strongly absorbed, while the adjacent sections either remain bright, or the absorption is much less vigorous. With certain dilutions, there will appear one or more narrow or broader absorption-bands, the position and extent of which are definable by means of the scale in relation to the Fraunhofer's lines.

The position of absorption-bands is either defined simply by the corresponding divisions of the graduated scale, adding, if necessary, the Fraunhofer's lines between which they are located; for instance a certain band may be said to have the position of 53-57, between D and E.

Or the following method introduced by Valentin may

* Gladstone, J. H., on the use of the prism in qualitative analysis. Chem. Soc. Journal, 1858. (Absorption Spectra of various colored metallic salt solutions.)

be adopted. Suppose there be 19 divisions between D and E, a band may be thus described :

$$D\tfrac{2}{19} E \text{ to } D\tfrac{7}{19} E.$$

A band commencing at A, and extending midway between B and C, is described as a band from A to $B\tfrac{1}{2}C$. The location of bands by this latter method is verified with greater facility by other observers than the determination by the former plan, which would require calculations similar to those referring to the Fraunhofer's lines (p. 10).

These absorption spectra must not be estimated at the same value as the flame spectra of the metals. The most valuable element of these latter is, that in a mixture of metal compounds, the individual metals may be discovered and distinguished with so great a facility; and this can scarcely ever be attained in regard to pigment mixtures.

Further, the absorption phenomena, consisting merely in the weakening or entire extinction of several parts of the spectrum, will rarely offer the sharp outlines of the bright metal lines,—the transitions from the absorbed to the unaltered part of the spectrum (the limit of absorption) being seldom definable with precision. These appearances are also not only dependent upon the pigment itself, but upon a number of other incidents, as the thickness of the layer of the fluid, the degree of concentration, temperature, light-intensity of the spectrum, admission or exclusion of air, etc.

We now propose to discuss the Spectrum appearances of the colored fluid, which preëminently claims our attention.

A valuable generalization is contained in the following sentence :

" Each body appears dark in the light, which it is incompetent to transmit." Applying this to the blood-pigment,* we may say:

*Hæmoglobin* (Hb)†, when combined with oxygen, appears dark at two distinct points of the yellow-green, between the Fraunhofer's lines D and E of the sun-spectrum.

Hoppe-Seyler, in 1862, taught this fact,[a] in regard both to the *Oxyhæmoglobin* [Scarlet Cruorin, Stokes] crystals ($O_2$-Hb), and to their solutions—and also to blood-solutions.

[a] Virchow's Archiv 23, 1862. Hoppe-Seyler (first communication).
" " 29, 1864. " " (2d and 3d communication).

The most important paper on the subject of the physiologo-chemical (including spectrum) relations of Hæmoglobin and its products of decomposition is: Hoppe-Seyler, Beiträge zur Kenntniss des Blutes des Menschen und der Wirbelthiere, in Med. chem. Untersuchungen. Berlin, 1866–1871.

He describes the gradual development of the $O_2$-Hb spectrum in about the following way:

A concentrated Hæmoglobin solution allows but a portion of the red to pass through. Diluted with water, it shows, when examined by sunlight, a rapid clearing up to D; light speedily appears between E and F, in the green; and, after further dilution, the spectrum expands over F, while, at the same time, a bright yellow-green band is seen midway between D and E, leaving two dark bands in this space. On continued dilution, the entire spectrum will gradually appear up to the violet, and the two absorption-bands alone remain distinctly visible at a dilution of 1 grm.

* Kuehne found the red pigment of muscle to be identical with Hæmoglobin (independent of the blood therein contained). E. Ray Lankester (Pflueger's Archiv iv. 1871) confirmed this in reference to particular muscles of various animals, especially the muscular fibres of the pharynx in certain mollusks, the blood of which does not contain any Hb.

† Syn: Hæmatoglobulin; Cruorin (Stokes).

Hæmoglobin to 10,000 C.C. fluid, and a thickness of the solution of 1 Cm.*

The first band ($\alpha$), nearer to D, is darker, narrower, and more sharply defined than the second ($\beta$), which nearly approaches E; and upon increased dilution, the first band will at last disappear, but somewhat later than the second band.

$O_2$-Hb is deprived of its oxygen by setting aside a sufficiently diluted blood-solution in a hermetically-sealed vial, or by raising such a solution, in a closed vessel, for a short period, to the temperature of the body,—or by adding reducing agents, for instance, one or several drops of liquid hydrosulphate of ammonia. The arterial color will then gradually fade, and the examination per spectrum present the bright interval between the two bands obscured, the bands themselves becoming fainter, and being replaced at last by the so-called *reduction-band*, a broad, imperfectly-defined absorption-band,—the darkest field of which is central between D and E. At the same time the blue part of the spectrum appears much less absorbed than in an $O_2$-Hb solution of equal concentration. If such a solution of reduced Hb (purple Cruorin, Stokes) be shaken up only for a short time with oxygen or air, the two bands will instantly reappear, to vanish and again give place to the reduction-band, if and so long as the reducing substance be present in the liquid. The reduction-band was discovered in 1864, by Stokes.†

Besides hydrosulphate of ammonia, other substances, which have a marked faculty of reduction in alkaline

* Of fresh defibrinated blood the proper dilution is 1 in 100 (vol.) water, to be viewed in a layer of 1 centimetre.

† George Gabriel Stokes, the celebrated occupant of Newton's chair, in Cambridge. (Philosophical Magazine, 1864.)

liquids, are employed. Among them are an ammoniacal solution of the tartrate of protoxide of iron,—an ammon. sol. of tartrate of protoxide of tin,—and sulphide of sodium (SNa), the latter warmly recommended by Preyer.* Of these reagents, the Tin solution offers certain special advantages. It is inodorous, colorless, and effects (even when added in larger quantity) simply reduction, while sulphide of ammonium and SNa decompose the blood, when employed in excess (Nawrocki). A blood-solution treated by protoxide of tin may be kept for weeks in open vessels, without undergoing the least change.

The tin-liquid is thus prepared: An aqueous solution of the common protochloride of tin (tin-salt) is treated by tartaric acid and neutralized by ammonia. The acid has to be added in such a quantity, that, after over-saturation by ammonia, no precipitate results, leaving a perfectly clear solution.

We are thus enabled to distinguish through the spectrum the presence and absence of oxygen in blood, because the blood is in each case of a different color. Alterations of the blood to be discernible by absorption-spectrum analysis, must include the formation of new pigments, which need scarcely be discernible to the unaided eye.

It will suffice here to refer to the close analogy between the artificial reduction and the change which the arterial blood undergoes in the living organism in its transformation into venous blood.

Beside the reduction, we have the decomposition of the bloodred, which by its products becomes the object of Spectrum Analysis. Of these products (albumines, pigments, or acids), the pigments alone affect the present question.

* Preyer, W., Die Blutkrystalle, Jena, 1871.

Several of these are artificial products of decomposition, which also occur in the living body. The most important of this class is *Hæmatin.*

Hb being a colored albuminous compound, nothing is required to effect its decomposition but the means ordinarily used for coagulating or precipitating albumen; viz., heat, alcohol, mineral acids, organic acids (even the weakest), and strong alkalies.

*Hæmatin* (Htn) being easily soluble in alkaline solutions *or* in acidulated alcohol, shows a different color in both. The alkaline solutions are of a reddish hue (Stokes's red hæmatin) in thick—green in thin layers; the acid solutions are brown (Stokes's brown hæmatin) in any thickness of the layer.

In each of these solutions, the least absorption takes place in the extreme red up to the line B; and both present, if sufficiently diluted, a characteristic absorption-band in the red, between C and D. But the very faint, broad band of the alkaline solution (the Htn band) is located near D (between C and D)—while the narrower and dark band of Htn dissolved in alcohol and $SO_3$ is situated close by C, between C and D (band of Htn in acid solution, Hoppe). This solution also shows frequently three other bands, two at the situation of the $O_2$-Hb bands (differing from them as to breadth and intensity) and one between *b* and F (*Hæmatoin* Spectrum, Preyer).

Valentin * was the first to distinguish the two bands—the first, the alkaline band as the "Htn-band,"—the other (as produced by concentrated acetic acid), the "*Hæmin*-band." Preyer calls the first one "Oxygen-Alkali-Hæmatin"; and the second "Acid-band."

* Loc. cit. pp. 80 and 82.

Valentin's Hæmin-band, and Preyer's acid-band have the same position, viz.: between C and D, near C, extending beyond, and being nearly bisected by this Fraunhofer's line.

Though Hoppe's band of Htn in acid solution has not quite the same position, it is nevertheless the same, as the situation of these so-called "acid-bands" varies with the kind of acid used, and even with the quantity employed. Their position can only be generally determined as between B and D. The displacement of the band towards A is in direct proportion to the quantity of the acid used (Preyer).

A Htn solution suitable for Spectrum Analysis may be prepared, (according to Valentin, after the most essential features of Wittich's method) in this manner: Blood is shaken up with a solution containing one part of carbonate of potassium and two of water (in the proportion of one part of blood to 6 or 8 of solution), and subsequently filtered. Of the chocolate-colored precipitate, a portion is boiled with aqueous alcohol.

A "Hæmin" solution (Hm) for spectroscopic purposes is prepared by heating a blood-solution, to which acid. acet. concent. had been added. From this Hm solution a fluid presenting the Htn spectrum can be readily prepared by adding ammonia to the former.

Htn (containing O) is also reducible by any of the above-quoted reducing agents. After adding any of them to a Htn solution, the fluid changes its color, and two bands, between D and E, the bands of *reduced Htn* are presented in place of the one.

They are easily distinguished from the normal bands,—the first normal band being very near D, while the first band of reduced Htn is so far distant from D, that a considerable part of the green-yellow is distinctly observable between it and D. Moreover, this band is made especially prominent by its great intensity and darkness. The second, a very faint band, includes the line E, and extends to or over *b*.

According to Hoppe, the spectrum bands of the reduced Htn (Stokes) are identical and belong to the alkaline solution of *Hæmochromogen* (a primary product of decomposition of Hb by acids or alkalies). Hence Hoppe calls Htn the product of oxidation of Hæmochromogen.

The second product in order of importance is the so-called *Methæmoglobin*, an intermediate result of the spontaneous decomposition of Hb into Htn and albuminous matter. The band, by which it is characterized, is situated between C and D, nearer C. It appears as soon as a decomposition of Hb sets in; and, according to Preyer, even when a pure Hb solution is set aside for a couple of hours, at the ordinary temperature. Sorby remarks that it is found also in the living body, in the dried blood-scabs of wounds in the process of healing. Some authors identify it with the acid-band. But, by bringing a little acetic acid into a solution of Methb before the slit of the spectroscope, the difference may be readily perceived (Preyer), a displacement of the band towards B taking place instantly. Nevertheless, it is produced by one of the weakest acids, $CO_2$ (Lankester).

We close this chapter with a note on *Hæmatoidin* and *Lutein*.

Hæmatoidin crystals (identical with Bilirubin, according to Hoppe) give two bands; one strong band between *b* and F, very near F, and a fainter one between F and G.

Lutein (Thudichum,* 1869), the yellow pigment of the

* Thudichum, Lutein. Centralblatt, 1869. Also Report of the Medical Officer of the Privy Council. London, 1870.

corpora lutea of mammal ovaries, and of yolk of egg, absorbs in concentrated solution the blue and violet light strongly.

When diluted before the slit by alcohol or ether, it presents two absorption-bands, one including the line F, extending more towards G, the other midway between F and G.

# CHAPTER IV.

## PRACTICAL APPLICATION OF THE BLOOD SPECTRUM.

THE Blood spectrum is employed, in the first place, for the *qualitative* demonstration of Hb, for the purpose of recognizing blood in secretions (urine), transudations, or in pathological products in general. And further, in Forensic Medicine, in suspicious stains.

Secondly, for the *quantitative* determination of the blood-red in the blood of an individual (in physiological or pathological cases), and of the quantity of blood in general.

In order to decide the presence of blood in urine by the Spectrum test, the urine has first to be sufficiently diluted and filtered. In the fresh urine, within the first 24 hours, the normal bands will be perceived, if it contain undecomposed blood; if the blood is decomposed, a hæmatin band will appear. The latter will be presented either without any special treatment of the urine, or it may become necessary to previously boil the urine and treat the coagulum obtained, with acidulated ($SO_3$) alcohol, gently heating it. In instances where no blood-corpuscles can be detected by the microscope in a blood-colored or ink-black urine (as in scurvy, putrid, typhoid, and malignant fevers, or after inhalation of $AsH_3$), the spectrum test will be specially important.

When blood is intermixed with vomited matter, it will

certainly appear undecomposed in the case of profuse hemorrhage from the stomach; but the so-called coffee-ground matter contains only hæmatin. To separate it from foreign pigments derived from food or medicine, the fluid must be warmed with a little diluted nitric acid, filtered, and the residuum dissolved in diluted caustic soda. The faint Htn band is then perceived before the slit, but it will be found better to change it by a reducing agent, into the two bands of reduced Htn.

The spectral relations of blood, extravasated into the gall-bladder, are analogous to those of blood admixed with urine. It will there appear unaltered only in the absence of bile: For bile dissolves not only at the temperature of the body the blood-corpuscles, but it transforms Hb into Htn and albuminous matter. Hb, changed in such a manner, is occasionally found in the gall-bladder in the form of crumbled precipitates, which dissolved in a little diluted caustic soda, will give the spectrum of Htn.

Hæmatin is a frequent constituent of fæces, the pigment originating in the normal intestine from animal food, but appearing also whenever an hemorrhage takes place in any part of the intestine, the lower section of the colon excepted; provided the functions of the intestine be not entirely disturbed, as in cholera, where the blood corpuscles are found unaltered in the dejections. For the purpose of demonstrating Htn in the fæces, an extract has to be made with cold alcohol, the residue boiled with alcohol, and a few drops of $SO_3$ added. This extract condensed by evaporation is then examined in the spectrum. (Hoppe-Seyler, Handbuch der physiol. und pathol. Chem. Analyse, 4th ed., 1875.)

Pathological occurrence of Hb or Htn in serous fluids is easily demonstrable by the Spectroscope.

In forensic demonstration,* the object under examination (the stain) should be first dissolved in ammonia water. This not only liquefies dry blood with greater facility than pure water would, but at the same time gives a brighter color to the solution, and will make the absorption-bands far more distinct, an advantage which will be especially appreciated in the examination of very minute quantities of the pigment. Of course the solutions must not be turbid. If they are so from dust or insoluble albuminous matter, precipitated through the spontaneous decomposition of the blood, they ought to be filtered. A deficient concentration may be compensated by making the observation through a thicker layer, or using a greater intensity of the transmitted light.

If we in such a manner succeed in discovering the $O_2$-Hb bands, there is scarcely a doubt about the presence of blood. However, the reduction-test will have to be made, since certain liquids present similar bands. A diluted ammoniacal solution of carmine, for instance, yields two bands in a similar position, but which may be readily distinguished from the blood-bands, the first being the fainter, and the second the darker one, in the carmine spectrum. The position is, besides, not at all identical with that of the blood-bands, as will be seen by comparing the spectra 16 and 17 in the table. The reduction-test, would, of course, have no effect upon the carmine spectrum.

Again, a solution of aniline red gives a single band (Table, Fig. 10), similar to the one band of reduced Hb ; it will be well, therefore, to carry O to the blood after the reduction, in order to restore the $O_2$-Hb bands, which pro-

* Hoffman, E., Professor at Insbruck. Einiges ueber forensische Untersuchungen von Blutspuren. Eulenberg's Viertelj. 1873.

cedure would not produce any change in the case of the aniline solution.

Other reactions, as Htn or Hm formation, may evidently be applied to the diagnosis of blood in forensic cases.

Concentrated solutions from less recent blood stains will frequently present, besides the ordinary bands, the Methæmoglobin-band, which in a strong dilution, fades and ultimately disappears, while the $O_2$-Hb bands still remain distinctly visible. Nevertheless, the appearance of this band is not applicable to the determination of the age of a blood-trace in forensic cases. E. Hoffman (loc. cit. p. 134) has shown by experiments, that the existence of Mthb in a blood-solution by no means allows the conclusion that the blood stain in question is of a remote date; and that on the other hand, the absence of this band does not furnish absolute proof that the blood under examination is of more recent origin.

The spectroscopic examination of blood stains will be of especial value in those instances where the microscopic demonstration of the form-elements of blood (particularly the corpuscles) is prevented through the action of water (water colored bloody by washing hands or clothing covered with blood).

As to the general value of the method, it must be remarked that the non-appearance of the blood-bands by no means justifies the conclusion that blood is absent.

In the first place, the solution may have been mixed with some other pigment by which the blood-bands might be obscured. Further, we know by Valentin's researches,* that putrescent old blood may occasionally be met, present-

* Valentin, G., Histologische und physiologische Studien. Ztschr. f. Biologie, vi. 1870.

ing no trace of bands, whether diluted by a weak solution of caustic, or carbonate of potash, or by a stronger one of Iodide of Potassium. Proper reagents may, however, even in such solutions, produce characteristic bands (sulphide of ammonium produces according to Valentin, in such blood *two* reduction-bands instead one,—the bands of reduced Htn?), or even the ordinary bands may be produced by introducing nitric oxide gas ($NO_2$), *i. e.* bands identical with the normal bands as to position (see page 47).

We have so far spoken only of solutions. Now, the blood may certainly be brought immediately before the slit in thin layers, dried on glass or other transparent substances. But in this case one would be hindered in the application of reducing agents or other reagents. If, however, the material at disposal be very small, so that a solution of proper quantity and concentration be not attainable, the procedure recommended by Hoppe-Seyler (Handbuch etc.) may be followed. The aqueous solution of the stain, freed from all contaminations, is dried in a watch-glass protected from dust, and with the watch-glass brought directly before the slit. On account of the fissures forming after awhile, and disturbing the vision, it is advisable to make the examination while the stain is yet wet.

Since Hb in dried blood is gradually transformed into Mthb, which, in a thin layer, will not easily present its absorption-band, it may happen that the *unaided* eye at least may not discover any trace of blood bands, if the last method be employed (Hoppe).

For the *quantitative* determination of the blood-red in any blood for physiological or pathological purposes, a Colorimetric method has been applied. It consists in comparing the intensity of the color of the blood, diluted by water, with the color of a Hb solution of known concentration. This method is pronounced unreliable by Preyer,* who in 1866 substituted his spectrum method, which is based upon the fact that concentrated Hb solutions are, in a certain thickness of the layer, impervious to all rays except the red (even with strong light), while less concentrated solutions of the same thickness will leave unabsorbed a part of the green, besides red and orange. Thus by diluting with water a measured blood-quantum before the slit, until the green appears in the spectrum, the percentage of Hb in such blood can be determined, provided the amount of Hb be known for a test solution, which is (under the same circumstances) *just* pervious for the green.

The amount of Hb in the blood of the sick, has been made the subject of investigation by H. Quincke † (according to Preyer's method). Quincke has modified that method by making use of a hollow glass prism instead of the plane-parallel vessel, so as to vary the thickness of the layer, the dilution remaining unaltered.

He has compared the amount of Hb in the blood of patients suffering from Chlorosis, Leukæmia, Nephritis, Diabetes, Pyæmia, Ileotyphus, etc., with that of normal blood, and found, (the latter being assumed as = 1) a very considerable diminution in Chlorosis (0.36), Leukæmia lienalis (0.39), Nephritis (0.58); an increase, on the contrary, in a case of Diabetes (1.10)—results, which, of course, have only an individual signification.‡

* Annalen der Chemie und Pharmacie, 1866; and "Die Blutkrystalle," 1871.

† Quincke, Ueber den Hæmoglobingehalt in Krankheiten. Virchow's Archiv. 54, 1872.

‡ Compare also Subbotin, V., Mittheilung ueber den Einfluss der Nahrung auf den Hæmoglobingehalt des Blutes. Ztschr. f. Biologie, VII. 1871. And—Coze et Feltz, Recherches expérimentales. Gaz. Méd. de Strasbourg, 1869.

We cannot omit a reference to Vierordt's researches, who, since 1871, has made the quantitative spectrum determination of coloring matter in general the subject of very elaborate investigations [Berichte der Deutschen Chem. Gesellschaft, IV., and: Vierordt, die Anwendung des Spectral-Apparates zur Photometric der Absorptions-spectren und zur quantitativen chemischen Analyse, Tübingen, 1873]. Originally his method was designed for measuring the absorption of light by diaphanous colored media, in any special region of the spectrum; but the same may be employed for determining the amount of pigment contained in any solution. Without going into minute details, we may mention two of the various modifications of his spectrum apparatus. First, the division of the movable slit plate into two portions, each provided with a graduated delicate micrometer screw for reading off the exact breadth of the slit. Second, a contrivance in the ocular of the observing telescope, allowing the obliteration of all parts of the spectrum except that which is to be examined. The transparent bodies are brought before the upper half of the slit. The intensity of the respective region of the spectrum has to be compared with that of the direct spectrum incident through the lower slit. This latter is narrowed gradually until the intensity of both spectra becomes equal, and in this manner the ratio of the widths of the two slits furnishes a measure for the amount of absorption of the medium under examination.

In conclusion, we have to mention a fact which may ultimately be of importance with regard to the examination of the blood of the sick; namely, the possibility of perceiving the absorption phenomena of the blood in the living organism. According to Hoppe the normal bands are visible in the ear of man, of white rabbits, the edges of the fingers held adjacent, the palm of the hand, and the natatory membrane of frogs.

According to Vierordt [Physiologische Spectral-Analysen. Ztschr. f. Biologie, 1875], the fourth and fifth fingers are brought into contact, so as to allow the sunlight to be visible only through the soft parts; the boundary-line of both fingers will then appear of a much brighter red than the scarcely translucent phalanges. This boundary-line must be laid upon the slit, when the bands will be perceived and may be determined. Vierordt saw even the reduction-band appearing in the spectrum, after putting rubber rings around the first phalanges of both fingers, compressing thereby the soft parts sufficiently to cause an interruption of the circulation.

## CHAPTER V.

### ON THE ABSORPTION–BANDS OF BILE AND URINE.

FRESH ox-bile has a green color, and shows in the spectrum, if the layer be rather thick, a band between D and E, nearer D (Hoppe).

Fel tauri inspissatum, treated with muriatic acid, presents this band very distinctly.

The four best known bile-pigments, Bilirubin, Biliverdin, Bilifuscin, and Biliprasin, yield (only after treatment by nitric or muriatic acid [oxidation]) a pigment, characterized by four absorption-bands—the first appearing at C, the second before and close to D, the third just behind D, and the fourth near to and before E. This pigment (also obtained from ox-bile, after it has been left standing), has been called Choleverdin by Stokvis, and Bilicyanin by Heynsius and Campbell.*

Our more intimate knowledge of the spectrum appearances of bile and urine dates from Max Jaffé's researches,† 1868. He made the interesting discovery that the well known Gmelin's test for bile-pigment (by nitric acid, containing nitrous acid), which presents a scale of colors (green, blue, indigo, red, and yellow), causes characteristic changes

* Heynsius und Campbell, die Oxydationsprodukte der Gallenfarbstoffe und ihre Absorptionsstreifen. Pflueger's Archiv f. Physiologie, 1871.

† Jaffé, M., Beitrag zur Kenntniss der Gallen—und Harnpigmente. Centralblatt. 1868, No. 16.

Jaffé, M., Zur Lehre von den Eigenschaften und der Abstammung der Harnpigmente. Virchow's Archiv. 47, p. 405.

in the spectrum. As soon as the color of the solution approaches the blue modification, a broad dark absorption-band appears between C and D, commencing nearer D, and extending midway between D and E. On dilution, this band separates into two indistinctly circumscribed bands ($\alpha$ and $\beta$), with a brighter interval at D. These remain visible, though with decreasing intensity, until the appearance of the red modification. Almost simultaneously with $\alpha$ and $\beta$ (but generally somewhat later), between *b* and F, almost exactly bounded by F, a third band ($\gamma$) becomes visible, which is proportionately more distinct as $\alpha$ and $\beta$ fade away.

The pigments corresponding to these bands, which appear only in the acid solution, have been isolated by Jaffé.

By alkalizing this acid solution, a fourth band is produced.

The reddish filtrate of an extract of human or dog's bile (this extract being made by means of diluted muriatic acid) shows in proper dilution the band $\gamma$ much stronger and more sharply defined. This solution alkalized by caustic soda turns bright yellow, $\gamma$ vanishes, and, after a few minutes, a new band ($\delta$) appears, likewise between *b* and F, but nearer *b*. It is less distinct, if ammonia be used instead of soda. Both bands may appear together, if the alkali has been added in slightly larger quantity; a trace of acid or alkali would then alternately suppress the one or the other.

The same pigment, giving the three absorption-bands, is found preformed, according to Jaffé, in gallstones.

It is a remarkable fact that the pigment produced by diluted muriatic acid from bile is identical with one of the normal urine, called by Jaffé *Urobilin*. Normal human urine, without chemical treatment, frequently presents—though indistinctly—the band $\gamma$, when the layer is sufficiently thick (3–6 Cm), and examined in a strong light. It will be much more distinct in the normal urine, if the latter (1–200 Ccm) be precipitated by acetate of lead, the precipitate decomposed by diluted sulphuric or oxalic acid, and the reddish yellow filtrate examined. Upon adding caustic soda, the displacement of the band towards *b* (the band *δ*) is then also perceived.

But all these spectrum characteristics may be, with the greatest facility, perceived in urine of greater concentration; viz., in the bright red urine of fever patients, examined without previous preparation in a test-tube. The spectrum appearances are not destroyed by boiling the urine with diluted mineral acids or alkalies; even when set aside for weeks, and having become completely musty, it will present the bands, so long as it has not turned putrid,—which proves that the coloring matter of the urine is not so easily decomposable as generally thought to be.

Urobilin is not only distinguished by characteristic spectroscopic qualities, but by brilliant fluorescent phenomena, which are produced with facility by treating urine with ammonia and ClZn.—Jaffé found further, that fresh pale urine, which would not show a trace of an absorption-band, will often become darker after remaining exposed to the air, and will present the band characteristic of Urobilin.

This proves (according to Jaffé and Hoppe) that in normal urine, a "chromogen" of Urobilin exists; which as Jaffé has shown by experiments, is transformed into the pigment by taking up oxygen.

Closely connected with the whole subject is a fact of great scientific importance, established by Hoppe [Hoppe, Einfache Darstellung von Harnfarbstoff aus Blutfarbstoff. Berichte der Deutschen Chem. Gesellschaft, 1874.] At a prior date, a reduction-product of Htn, obtained by treating

Htn in alcoholic solution with Tin and ClH, had been described by Hoppe, and supposed by him to be identical with Jaffé's Urobilin (= Maly's Hydrobilirubin); but recently he found the same pigment to be obtainable with equal facility by treating undecomposed Hb in alcoholic solution with Tin and ClH. Hence, the coloring matter of the normal fæces and urine may be conceived as a product of decomposition of the blood-pigment, altered by reduction.

Bilirubin and Biliverdin also represent intermediate stages of this metamorphosis, or they are, at least, closely related to blood-pigment.

Though similar views have long been held, still they have only attained a secure basis through Hoppe's discovery.

Hoppe remarks, at the close of the paper we have quoted: "Physiological questions of great importance are hereby presented. For instance, it will be possible to find how great, under normal or pathological relations, the decay of blood-pigment or of red blood corpuscles will be in a definite time."

Pettenkofer's well known test for bile-acids (by cane-sugar and $SO_3$) is liable to disturbance by the presence of albumen, since albuminous matter treated by $SO_3$ yields a similar purple color. By the use of the spectroscope, the absence of albumen may be established. According to Schenk (Hoppe, Handbuch, etc., 1875, p. 104), the purple solution of bile acids treated by $SO_3$ and sugar, presents in proper dilution with alcohol, a band between D and E, next to E; and a second band before F, which does not take place, if albumen be treated in the same manner.*

* The spectral relations of the bile-pigments are thoroughly discussed in Prof. J. C. Dalton, Human Physiology, 6th edition, Philadelphia, 1875.

The partial identity of the spectrum ($\gamma$) of bile-pigment and Urobilin (if the latter be considered as a normal coloring matter of the urine), will, it appears, prevent a ready application of the spectroscope for discovering bile-pigment in icteric urine.

Stokvis (1870) treated icteric urine by ClZn and ammonia in excess, and found three bands: one behind C, another at D, and a third between *b* and F. It was only in icteric urine that he obtained the two first; the third was also found in normal urine.

As a product of decomposition of Indican, Indigo occurs in putrescent urine, for the spectroscopic diagnosis of which the absorption appearances of Sulphindigotic acid (Indigo dissolved in conc. $SO_3$), may be employed. It shows an absorption between C and D, in thicker layers extending over D. (See Table, No. 7.)

## CHAPTER VI.

### ON SPECTRUM ANALYSIS OF THE BLOOD AFTER INTRODUCTION OF FOREIGN GASES.

THE modifications of the normal blood-spectrum, after the introduction (or inhalation) of foreign gases, are of course based upon the various alterations which the Hb thereby undergoes.

Certain gases expel O from its combination with Hb; and, supplanting it, form new compounds, of greater stability, with Hb; thus preventing it for a longer or shorter term, from recombining with O.

Others have the property of reducing $O_2$-Hb, when introduced for a protracted period, even at a low temperature, but do not combine with Hb, which will again take up O on shaking the solution with air. To this class belong Hydrogen and Nitrogen.

Certain gases have not only a reducing but a decomposing effect; and, lastly, there are gases by which the Hb will be completely destroyed.

Before discussing these spectrum appearances in detail, we have first to make some remarks on the special methods adopted in such investigations.

In general there are three methods employed in these researches:

1st. By direct introduction of the gases into Hb–or blood solutions.

2d. By experimenting with the gases upon animals.

3d. By observations of cases of gas-poisoning in man.

This third method has been applied, so far, only in reference to the action of two gases, CO and CyH.

For a number of the other gases, experiments have been instituted in accordance with the first and second methods.

Setting aside the fact that conclusions deduced from experiments upon animals can only be accepted with the greatest caution as bearing upon man, these experiments present so striking a contrast in their results, compared with those obtained from direct introduction into blood-solutions, that for the time we shall be scarcely tempted into drawing such conclusions.

For instance, we find in Hirt's "Gasinhalations-Krankheiten der Arbeiter," 1873, that, while $CO_2$ directly introduced into blood produces the reduction-band, the blood of animals killed by inhalation of the same gas only present the normal bands, which may subsequently be reduced *ad libitum*.

The same contrast is found in all comments on spectrum appearances as to HS, $SO_2$, and $NH_3$; that is, in reference to all those gases in which experiments, after either method, have ever yet been made (CO and CyH being here excluded from discussion).

Now we are certainly taught by physiological chemistry (vide Kühne, Lehrbuch der physiol. Chemie, p. 247) that "the alterations in the blood after poisoning are identical with those produced by the same poisons in blood freshly drawn from the body. This holds good especially for $CO_2$, CO, HS, for $AsH_3$, and $SbH_3$."

But, remarkable as the contrast pointed out may be, we must refrain from concluding, *a priori*, that it is impossible.

We shall see, on the contrary, that not only have good and valid reasons been brought forward for such a difference in action by the most eminent men of science, but that an attempt to controvert these results would, for the time, have but little effect.

It has formerly been the practice in experimenting upon animals, to examine the diluted blood without excluding the atmospheric air. Hb having a great affinity for O, will combine with it, through mere dilution by water, and will thus yield the normal bands, unless made incapable of absorbing O, by such a gas as CO.

The exclusion of air in these examinations was entirely unknown before 1867. We are indebted for this method to Prof. Gwosdew,* (Krakow) who conceived the idea of applying Spectrum Analysis to the forensic diagnosis of death from suffocation. The fact of the absence of oxygen in the blood of suffocated animals had been previously made known by Setschenow; that the air contained in the lungs of the suffocated animal was deprived of O, had also been demonstrated by W. Müller. Gwosdew consequently expected to obtain the reduction-band in the blood of suffocated subjects, provided the atmospheric air could be completely excluded during the examination. By means of a certain procedure (described loc. cit.) he indeed succeeded, in several experiments, in obtaining the postulated band; whence he concluded death by suffocation to be diagnosticable by that means. The experiments were repeated by F. Falk,* (Berlin), with an apparatus of his own invention, with entirely negative results, due, as was shown

* Gwosdew, J., Spectroscopische Untersuchung des Blutes bei Erstickten. Reichert's Archiv. 1867.

* Falk, F., Verwendung des Spectroscops zu forensischen Zwecken. Prager Vierteljahrschr. 1869.

by Kotelewski,* (Warsaw) in 1870, to some defect in its construction, K. at the same time relating the positive results he had obtained by Gwosdew's method as well as his own.

His apparatus and method of procedure are as follows: A test-tube, narrowed and drawn out at the upper end, is filled (after warming it,) with pure glycerine, mixed with a little distilled water, to the level of *b*. Upon the open end a short rubber tube is fixed, and tightly bound with thread; the fluid is then subjected to a boiling heat, in order to expel the air from the mixture. The rubber tube is then drawn over the thick end of a metal tube, provided with a stop-cock, and at the opposite end with a pointed canula, the bore of the latter being somewhat wider than that of a common Pravaz syringe. The rubber tube is now bound over tightly with thread; the boiling of the glycerine continued cautiously for a little while, until the level of the fluid reaches about $b^1$; the canula-point is now immersed in a vessel filled with thoroughly boiled glycerine, and the stop-cock closed.

Then the glycerine, completely deprived of air, is allowed to cool.

After cooling, the apparatus is ready for the reception of the blood, the point being inserted in the heart or one of the larger veins; and, as soon as the stop-cock is opened, the blood rushes into the vacuum. The stop-cock is now closed, the apparatus shaken, and brought before the slit.

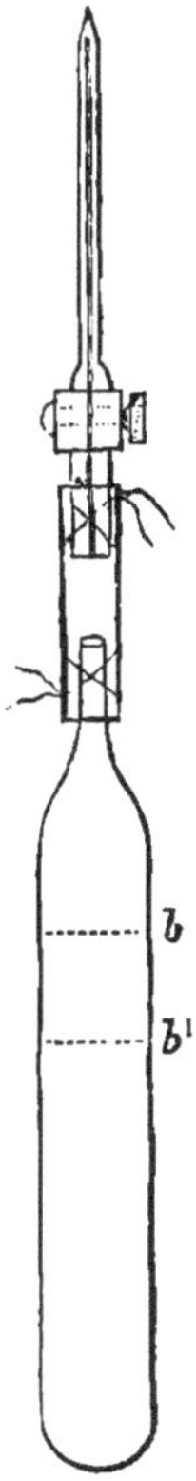

Fig. 4.

* Kotelewski, Zur Spectral-Analyse des Blutes. Centralblatt, 1870. No. 53 und 54.

Although obtaining the reduction-band in all his experiments, Kotelewski came to the conclusion, nevertheless, that Spectrum Analysis was not applicable to the diagnosis of death from suffocation, for the following reason: He had found the blood of every dead animal or human subject under examination (atmospheric air being excluded) to present the absorption-band of reduced Hb, and that it was absolutely impossible to obtain the $O_2$-Hb bands under those conditions. The tissues, he says, consume oxygen so rapidly as to leave only reduced Hb in the blood of the veins a few minutes after the lungs have ceased carrying O to the blood. He has not omitted to examine the blood from a vein of a living animal in the same way, and found it to contain $O_2$-Hb. All these remarkable observations of Kotelewski were subsequently confirmed by Falk, in a short paper published in the "Deutsche Klinik," 1872. It is therefore an established fact that Spectrum Analysis is decidedly inapplicable to the medico-legal diagnosis of death from mechanical suffocation. The subject is treated in this sense in Caspar's "Hand-book of Legal Medicine," edit. 1871.

There is no doubt but that such a mode of investigation in gas-poisoning will, in certain instances, give quite a different result from what has been formerly taught; though it will be apparent that the practical effect, as to a post-mortem diagnosis, will be as nugatory in poisoning by "reducing" gases, as it has been shown to be in mechanical suffocations. Nevertheless, it may become a very useful means of inquiry in experiments upon animals, as well as in observations upon man, if undertaken *before* death has set in. On the other hand, two bands equal in position to the normal bands, would, if obtained from a dead body (the blood being examined under exclusion of air,) certainly go far to prove a deep alteration of the Hb, especially if found irreducible. The reduction-test could not in such cases be made in the usual manner, unless the reagent could be admitted without the introduction of air.

Kotelewski's method might probably be supplanted by one which could be performed with greater facility; viz., the examination of the undiluted blood, in a thin layer, taken from the animal in an exhausted vessel or glass-tube.

We shall see, further on, how other incidents beside the admission of air, may possibly prove the cause of the striking incongruity we have pointed out.

We now proceed to discuss in detail the spectrum appearances of the blood with regard to various gases, or vapors, though some, such as bisulphide of carbon vapors, exhibit no influence whatever upon the normal spectrum.

The so-called indifferent gases, Hydrogen and Nitrogen, we have already alluded to.

Three gases, $NO_2$, CO, and CyH, are known to expel oxygen from its combination with Hb, supplanting the oxygen, and forming stronger compounds with Hb.

$NO_2$. Nitric oxide (binoxide of nitrogen) is prepared by treating copper clippings with nitric acid. (Deoxidation of $NO_5$ by the metal.)

After introduction of nitric oxide gas into $O_2$-Hb solutions, the normal bands appear unchanged. They no longer pertain to $O_2$-Hb, but to a newly-formed compound, discovered by Prof. L. Hermann,* and prepared by him in crystal-form, $NO_2$-Hb. Hermann introduced $NO_2$ (by

* Hermann, L., Ueber die Wirkungen des Stickstoffoxydgases auf das Blut. Reichert's Archiv. 1865.

exclusion of air) into a blood-solution, previously deprived of O by Hydrogen gas. The blood became of a bright crimson color, and presented, instead of the reduction-band, two bands equal in position to the $O_2$-Hb bands, but of a less intensity. These bands cannot be transformed by the addition of a reducing agent.

Since it can never be inhaled, being converted by the contact with air into the irrespirable Hyponitric acid ($NO_4$), the knowledge of its spectrum relations to blood has apparently no practical value. It is, however, spectroscopically interesting, in that it shows that bands almost entirely identical with the normal bands may have quite a different signification. It may also be used, as mentioned above, as a means of detecting the presence of blood in cases where, from its age, the original normal bands are entirely absent.

CO. This gas is readily prepared by heating oxalic acid in a flask with five or six times as much conc. $SO_3$. The acid being resolved into water, $CO_2$ and CO, the gases must be passed through several wash-bottles filled with a solution of caustic soda (half dilution) for the absorption of $CO_2$.

Pure carbonic oxide gas, introduced into $O_2$-Hb solutions, expels oxygen, supplanting it, and forming the crystallizable compound CO-Hb (Hoppe), known before $NO_2$-Hb. The bright cherry-red solution yields, instead of the normal bands, two bands almost identical to these, which may be distinguished by their displacement towards the violet end of the spectrum.* The difference of position being scarcely sufficient for their recognition, the reduction-test must here again decide their character.

Upon adding reducing agents (Tin solution in prefer-

* In observations of this kind it is important to know, that by widening the slit a perceptible displacement of bands is effected.

ence) the CO-Hb bands do not disappear, nor are they replaced by a reduction-band—at least, not instantly,—sometimes, especially at a low temperature, not before a lapse of several days. This remarkable condition exists also in the blood of animals killed by inhalation of the gas.

CO-Hb not being an inseparable compound, though of greater stability than $O_2$-Hb, it follows that, as Hoppe suggested, the examination must be made as early as possible after the poisoning has been discovered. With man, however, poisoning from pure CO will rarely occur; and it is evident that in those frequent cases of poisoning by CO-mixtures, such as charcoal fumes* or common gas, the spectroscopic appearances will be, from the presence of a large percentage of air, somewhat more complicated.

The bands presented by such blood, which does not always show the florid arterial color of CO blood, being frequently dark, may be displaced towards the more refrangible end of the spectrum; but on adding a reducing agent, a reduction-band will in such cases appear, due to the action upon the oxygen introduced with the fumes. That is, the space between the two bands will be obscured —but a more minute examination will show that the two bands have not disappeared. This, together with the changed situation (which may be verified by the simultaneous observation of a normal blood-solution—by means of the prism of comparison) will decide the character of the blood under examination.

This is not a mere theory, but the result of an observation † made by Prof. Hensen (Kiel) in a case of poisoning

* A valuable paper by A. Gamgee "On poisoning by carbonic oxide gas, and by charcoal fumes," is contained in the Journal of Anatomy and Physiology (London) I. 1867.

† As far as the author knows, the only detailed observation on record.

by charcoal fumes, (cured by transfusion,) communicated by Jürgensen. (Drei Fälle von Transfusion des Blutes, Berl. Klin. Wochenschrift, 1871, p. 304.)

The diagrams will illustrate Hensen's observation.

The spectra 1, 2, 3 are drawn according to Hoppe-Seyler;
" " 4, 5, 6 and 7 according to Hensen.

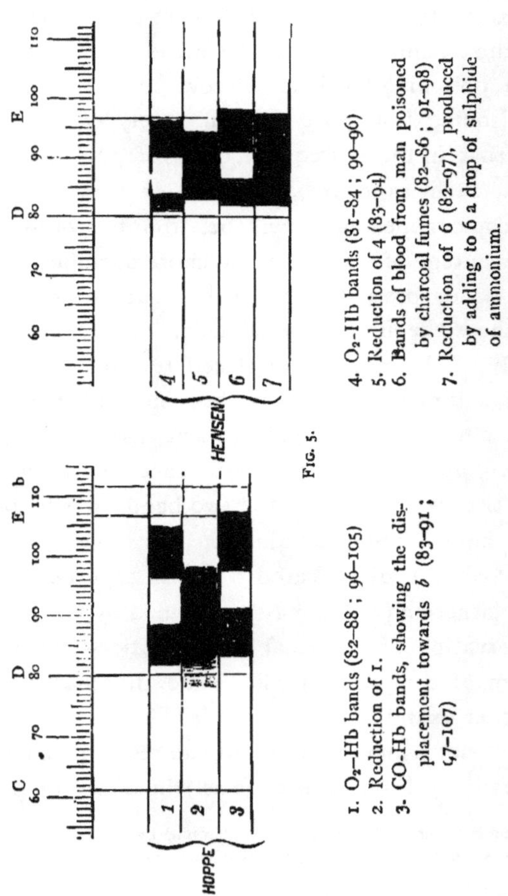


FIG. 5.

1. $O_2$-Hb bands (82–88 ; 96–105)
2. Reduction of 1.
3. CO-Hb bands, showing the displacement towards *b* (83–91 ; 97–107)

4. $O_2$-Hb bands (81–84 ; 90–96)
5. Reduction of 4 (83–94)
6. Bands of blood from man poisoned by charcoal fumes (82–86 ; 91–98)
7. Reduction of 6 (82–97), produced by adding to 6 a drop of sulphide of ammonium.

Though Hensen has made his observation without determining the Fraunhofer's lines, we are, from his description of the bands, enabled at least to locate his D at 80, and his E at a fraction beyond 96.

The bright interval between the two bands is obliterated (7) by a reduction-band (which would not take place in a pure CO-Hb solution); but the two bands are not destroyed. The contraction by one degree of the scale, observed at the right end of 7 was due, according to Hensen, to some $O_2$-Hb, which had contributed in his case to extend the second band (spectrum 6) towards *b*, and which had been reduced by the reagent.

CyH. Cyanide of hydrogen (hydrocyanic or prussic acid) also forms a strong and crystallizable compound with Hb (Hoppe, Preyer). The spectrum of a solution containing CyH-Hb does not present—according to Hoppe—any change of aspect from that of a pure $O_2$-Hb solution. On adding a reducing-agent, the two bands disappear and are replaced by the reduction-band. Nevertheless, this compound is more fixed than $O_2$-Hb. If, according to Hoppe, a blood-solution, to which a few drops of prussic acid have been added, be enclosed in a glass tube, so as to leave but very little air above the liquid, the mixture will present the normal bands, even after the lapse of several months, while a few days would suffice for the production of the reduction-band, if CyH had not been added.

Though Preyer's results differ considerably from Hoppe's, still observers agree in the fact that the gas cannot be spectroscopically demonstrated in the blood of poisoned animals. The same negative result appears in the single observation recorded concerning the spectrum examination of the blood of man after poisoning by prussic acid. H. Siegel (Leipzig) —"Fall von Vergiftung durch Blausäure," Archiv der Heilkunde, 9, 1868—found in this case, the blood dark, wholly fluid; in thin layers it appeared cherry-red, and the two absorption-bands of $O_2$-Hb presented themselves in all distinctness.

$CO_2$ is prepared by the decomposition of marble with diluted muriatic acid; or in greater purity, by treating chalk with concentrated sulphuric acid.

After the direct introduction of carbonic acid gas into a blood-solution, the normal bands disappear, giving way to the reduction-band. On shaking the solution with air, it again absorbs oxygen, the $O_2$-Hb bands re-appearing.

Nothing of this kind is seen in the blood of animals poisoned by $CO_2$. As we have previously remarked, only the normal bands are perceived, and these may be afterwards reduced by any reducing agent. An examination with the exclusion of air would doubtless give a different result. The questionable value of a reduction-band obtained under such conditions will be sufficiently demonstrated in the introductory remarks to this chapter.

Although usually nothing is mentioned about $CO_2$ but its reducing action, we should observe that it is not thus limited. On leaving a blood-solution into which $CO_2$ had been introduced, for several hours at a low temperature, without shaking it, the so-called Methæmoglobin band will appear. The same phenomenon will probably be perceived in the blood of animals poisoned by $CO_2$.

HS. After the direct introduction of sulphuretted hydrogen gas into a diluted, neutral blood-solution, the liquid turns to deep red; and besides the normal bands which soon become fainter, and disappear after a short time, being replaced by the reduction-band, the band appears, which is generally considered as characteristic and specific for HS blood (Hoppe-Seyler), and is therefore called the HS band. It is a well-defined absorption-band situated in the orange between C and D, about midway between both lines.

The reduction-band spoken of is, by shaking the solution with air, re-converted into the normal bands.

The chemical character of the pigment, represented by the HS band, is not yet thoroughly understood. The gas is oxidized in the blood to S and HO, and this is accomplished far more rapidly than it would be by the oxygen of the air, in the absence of blood; whence the conclusion that the ozonizing faculty of the blood pigment is performing an essential part in this process. Before the reduction of $O_2$-Hb, implied in this operation, has been completed, the pigment is formed which optically manifests itself by the HS band.

Hoppe concludes from its origin, that the body representing it is a sulphur compound, either of Htn or of Hb.

If the introduction of the gas is continued for a lengthened period, even this pigment will undergo decomposition, with the separation of sulphur, and an albuminous matter. At this stage of decomposition there are, however, no absorption-bands whatever visible.

That the HS band is not to be considered identical with the Htn band, as asserted by several authors, Hoppe has demonstrated in his paper*: "Ueber die Einwirkung des HS auf den Blutfarbstoff." The difference is not only

* Medic. Chem. Untersuchungen. Berlin, 1866.

in the position of this band and those produced from decomposition by acids or alkalies, but in the decisive point of the reduction-test. The non-identity of the HS and the Htn band is proved by Hoppe through the immutability of the HS band by reducing agents. While all Htn solutions, on adding sulphide of ammonium, soon exhibit their two reduction-bands in the green (the band in the red disappearing), the HS band will remain unchanged. The HS band differs also as to position from the Methæmoglobin band. It is besides remarkable that the latter will disappear after gently heating the fluid, while the HS band requires the application of boiling heat for this purpose (Preyer).

With a greater dispersion than that used for the spectrum table, the respective bands present the following positions:

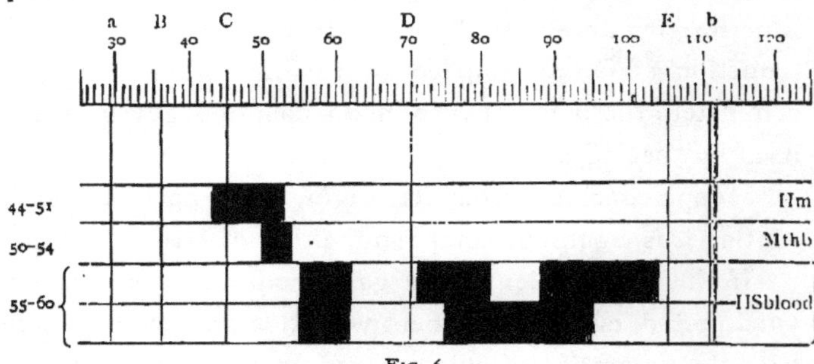


Fig. 6.

As to the two bands between D and E, which often remain visible in the spectrum beside the HS band, they are considered and designated by almost all authors as the normal $O_2$-Hb bands. Only Preyer ("Blutkrystalle," 1871) entertains the view—supported by arguments—that they do not pertain to $O_2$-Hb, but—as he says—probably to a combination of $O_2$-Hb with sulphur.

Now the HS band would certainly offer an excellent means of diagnosis in cases of poisoning by this gas. But here we have the oft-repeated contrast appearing strikingly.

Spectrum observations concerning the blood of man poisoned by this gas have as yet not been recorded. We are therefore limited to its results in experiments made upon animals. These records, however, afford us no evidence of the characteristics of this poisoning in the case of warm-blooded animals. Rosenthal and Kauffman,* and Hoppe have observed "the higher degrees of blood-decomposition" in cold-blooded animals, it is true; but concerning that class which more immediately interests us, they say "The normal bands are perceived; but a smaller quantity of sulphide of ammonium is required for withdrawing all O from the blood." At another place they say: "It seems quite self-evident that the primary effect of the gas introduced into the circulation must consist in withdrawing the O. That the other effects observed by us, the decomposition of Hb and Htn do not appear, is easily explainable, since death must set in, as soon as the larger proportion of O is withdrawn from the blood."

L. Hermann † is very explicit on the point in question. He says: "It is certain that, with warm-blooded animals, only the first stage of the described effects upon the blood can be brought about, as the reduction of the blood alone must, by necessity, cause immediate death from suffocation. There is, indeed, nothing to be seen in the blood of poisoned warm-blooded animals, of the absorption-band of the Hæma tin-like pigment above described."

* Rosenthal, J., und Kauffman, "Ueber die Wirkungen des HS Gases auf den thierischen Organismus." Reichert's Archiv. 1865.

† Hermann, L., Lehrbuch der experimentellen Toxicologie. Berlin, 1874.

Eulenberg,* who has made a number of experiments upon animals, makes no allusion to the point. Hirt (loc. cit. p. 47) follows Rosenthal and Kauffmann.

In all these spectrum observations, air had been admitted. In the only observation existing, made with exclusion of air, by Falk (communicated in the paper quoted above, Deutsche Klinik, 1872), the HS band is not alluded to.

But in Falk's observation we have another point worthy of special notice. He has obtained in his experiment the $O_2$-Hb bands, whence he concludes that animals poisoned by HS have still O in their blood, and that they are not therefore suffocated. To us this result appears in the light of Kotelewski's proposition, to offer almost a confirmation of Preyer's view, that the two bands do not appertain to unaltered $O_2$-Hb.

Thus we perceive that the two bands in the green do not at present offer anything characteristic for the HS spectrum; and the band in the red not having been found (before the slit) in the blood of warm-blooded animals, the contrast pointed out appears to be real also in this case.

But there yet remain some points of qualification. The HS band will, under certain circumstances, fail to appear even after direct introduction. Thus it is not seen, when the gas has been sparingly introduced, or in very diluted solutions. The latter peculiarity it has in common with other bands in the red. Furthermore, Valentin has, since 1863, repeatedly insisted upon the necessity of employing a bisulphide of carbon prism for bands of this description, especially for the HS band. Again HS being partially

* Eulenberg, H., die Lehre von den schädlichen und giftigen Gasen. Braunschweig, 1865.

eliminated through the lungs during the experiment, a difference in the spectroscopic results seems possible, according to the longer or shorter duration allotted to the procedure of poisoning. With this idea in view, the author has made a number of experiments upon rabbits, and has obtained the HS band with all desirable distinctness in two experiments, in which the animals were killed by the inhalation of the gas with the rapidity of lightning.

In seven experiments of a somewhat more protracted duration, nothing but the normal bands were perceived. In the first two cases the animals were placed under a narrow conical glass bell into which the gas had been freely introduced before; in the latter a large roomy vessel was substituted. The gas was developed in the usual manner by treating SFe with diluted $SO_3$, and conducted through a wash-bottle, before reaching the bell. The blood was taken without excluding the air, and diluted by water.

The prism employed was Eaton's compound $CS_2$ prism.* the apparatus being adjusted to the following scale:

| F | *b* | E | D | C | B | a | A |
|---|---|---|---|---|---|---|---|
| 28.8 | 55-56 | 63 | 100 | 127 | 136 | 144 | 154 |

The HS band seen immediately after the death of the animal, had the position 118—111, perfectly identical with the band obtained after direct introduction of the gas into a blood-solution, or after diluting blood with HS water, or after adding sulphide of ammonium in excess to a blood-solution. The two normal bands, appearing beside the HS band were found reducible.

It thus appears that, under certain conditions, the ex

* The band has been seen also by means of a simple flint prism.

periment upon animals with HS will give the same result, as the direct introduction into recently drawn blood.

That this observation, even if confirmed, will not be practically advantageous in forensic diagnosis, is evident. For, in the first place, the *pure* HS gas is probably never the cause of fatal accidents. Besides, we should, after the poisoning of *man* by HS mixtures, have reason to expect the characteristic band only under the conditions above described.

The value and importance of Spectrum Analysis does not depend upon its applicability to diagnosis in forensic medicine. The reader needs only to be reminded that HS, for instance, is employed even as a remedy, which fact is certainly sufficient reason why the physician should become acquainted with its spectrum deportment towards blood.

In connection with this subject we have to point out a possible source of error. In the very locality occupied by the HS band, there will appear, if sunlight be employed, under certain circumstances, a dark line belonging to the so-called terrestrial or atmospheric lines. This line will be widened by any blood-solution brought before the slit, and may thus deceive us. It is therefore absolutely necessary to make observations of this description not exclusively with solar light—but at the same time with artificial light—in which latter case no absorption lines will be shown.

$SO_2$. Sulphurous acid, for technical purposes, is usually prepared by the combustion of sulphur in an enclosed place, with a regulated admission of atmospheric air; and is obtained pure for physiological experiments by heating

$SO_3$ with copper clippings (deoxidation). Hirt gives (loc. cit. p. 71) the results of his valuable experiments made with this gas upon rabbits, dogs, and frogs. According to Hirt's researches the spectrum appearances are: After direct introduction of sulphurous acid into a blood-solution, the $O_2$-Hb bands disappear instantly, no other band appearing in their stead. If the air introduced into the blood-solution contained 30 per cent. $SO_2$, the bands vanish after half a minute; if 10 per cent., after 1½ minutes. The fluid at the same time assumes a peculiar brownish color, the Htn band, however, not being perceptible.

The blood of the animals killed by the inhalation of the gas did not present any perceptible change from the normal aspect.

$NO_4$ vapors. Fresh ox's blood treated by hyponitric acid vapors (Eulenberg, loc. cit. p. 247) coagulated rapidly to a chocolate-colored mass.

Gamgee* obtained an acid-band, after treating blood with $NO_3$ salts, the $O_2$-Hb bands simultaneously becoming very faint. After adding a few drops of $NH_3$, the blood was again restored from dirty-brown to its original red color, the new absorption-band disappearing, and the bands between D and E becoming much more distinct. The blood of animals poisoned by $NO_4$ (Eulenberg) has not been submitted to the spectrum test.

$NH_3$. Pure Ammonia is prepared by heating a mixture of Chloride of Ammonium and moist hydrate of lime in a distilling apparatus.

Koschlakoff and S. Bogomoloff,* in their experiments,

* Gamgee, A., Note on the Action of Nitric Oxide, Nitrous Acid, and Nitrites on Hæmoglobin. [Proceedings of the Royal Society of Edinburgh, VI. 1867.]

conveyed the gas through a rubber tube cooled by ice. The test-tube containing the blood-solution was cooled in the same manner.

Hirt (loc. cit. p. 93) also furnishes the results of his experiments.

Koschlakoff and B. remark that the solutions show primarily a yellowish tint, becoming gradually yellowish-brown, and at last brownish-green.

Hirt found the blood of animals killed by inhalation of a mixture of 90 per cent. $NH_3$ and 10 per cent. atmospheric air at first a dark red of a bluish tinge, rapidly brightening on contact with air.

All agree as to the effect upon the spectrum on direct introduction. The $O_2$-Hb bands gradually disappear, no reduction nor Htn band taking their place.

According to Hirt, the bands reappear after shaking the solution for a short time, vanishing after renewed introduction of the gas.

No abnormal condition, however, is perceived in the spectrum of the blood from animals killed by inhalation of the gas [*Hirt*].

$AsH_3$ and $SbH_3$. Koschlakoff and B. state that the introduction of arsenetted and antimonetted hydrogen also produces a rapid change of color. The solutions change to a yellowish-brown, afterwards to a greenish-brown. The $O_2$-Hb bands gradually disappear, and are substituted by the reduction-band, while the solution becomes slightly red. On the following day, the reduction-band vanishes in its turn, and traces of the normal bands are visible. The

* Koschlakoff und Bogomoloff. Ueber die Wirkung des Ammoniaks, des Arsen—und Antimonwasserstoffs auf die Blutpigmente. [Centralblatt, 1868. No 39.]

general assumption that these gases (especially $AsH_3$) destroy the $O_2$-Hb bands without reduction, must therefore be qualified; such a destruction taking place only in the thin layers.

$PH_3$. Phosphoretted hydrogen has, according to Koschlakoff and B., the same effect as $NH_3$ upon $O_2$-Hb. It destroys the bands without reduction or Htn formation. Dybkowsky,* however, obtained the reduction-band after conveying the gas through defibrinated dog's blood. On shaking the solution with air, the blood changed from the dark-brown color it had assumed to a somewhat lighter shade, when the normal bands re-appeared.

The blood of animals poisoned by $AsH_3$, $SbH_3$, and $PH_3$ has not been examined spectroscopically.

* Dybkowsky, W., Beitrag zur Theorie der Phosphor-Vergiftung. (Hoppe-Seyler's Med. Chem. Untersuchungen, I. 1866.)

# APPENDIX.

## NOTE ON THE EXAMINATION OF NARCOTIC POISONS AND REMEDIES.

VALENTIN * has examined with the spectroscope a great number of tinctures from narcotic remedies. None of them yields a spectrum sufficiently specific to enable us to distinguish them from each other.

He has further submitted to the spectrum test the bright-colored liquids, obtained by treating various poisonous alkaloids by sulphuric acid. To a small quantity of the alkaloid (Morphine, Narcotine, Strychnine, Brucin and Veratrine), he added from one to two cubic centimetres of strong sulphuric acid, containing $NO_5$; besides which he added a trace of $NO_5$ and small pieces of crystallized manganese. The vivid colors obtained by this procedure at ordinary temperature or after boiling, remained unaltered for three or four days, frequently for a longer period.

These researches were, however, equally negative in their result; in so far that in no instance were absorption-bands met with, which would be the only means of discriminating between the poisons.

Researches on the spectral influence of the admixture

* Valentin, Gebrauch d. Sp., etc., pp. 102 and 105.

of alkaloids to *blood* have been made within the last three years by Binz, and by Rossbach.*

Rossbach found that a blood-solution containing an alkaloid was equally well reduced by sulphide of ammonium as pure blood.

Nevertheless, it is a fact that the reduction of blood is delayed by alkaloids. Simple heating at from 40 to 50° C will reduce pure blood. Rossbach compared with that the effect of the admixture of an alkaloid; and found that, in this case, a higher temperature was required than with pure blood.

Binz has established the same retardation as to the reduction of $O_2$-Hb, in reference to Quinine.

Both authors connect the effect of the alkaloid with the transmission of ozone.

Rossbach infers that the alkaloids cause the ozone to combine more strongly with Hb, so that the latter will not be able to transmit ozone to other bodies so easily.

## NOTE ON THE USE OF THE SPECTROSCOPE IN OPHTHALMOLOGY.

IT is well known that the prism is largely used in Ophthalmology, for its *deviation* effects; but as our argument is confined to the *dispersion* of light, we shall notice only the application of the Spectroscope to cases of Color-blindness.

This affection has latterly engaged the attention of oculists in a much greater degree, on account of the practical importance it has assumed in reference to many

* Rossbach. Ueber die Einwirkung der Alcaloide auf die organischen Substrate des Thierkörpers. Verhandlungen der Physic. Med. Gesellschaft in Würzburg, III. 4, 1872; and VI. 3 und 4, 1874.

industrial occupations (such as railroad and steamboat employés, etc.).

An individual may be examined as to his reaction upon mixed or pure colors,* the former being those shown by an ordinary colored material, pigments, papers, etc., reflecting various colors, which together form the fundamental color; or, by colored glass, liquids, etc., which will transmit not one but several colors.

If, in a case of partial color-blindness, among a number of pigments, one is confounded with the others, this will have to be examined by prism or spectroscope.

In order to find the composition of an opaque pigment, we look through a prism at a thin stripe of the material, laid upon a dark ground (preferably upon black velvet). The light, emanating from the pigment, is thus decomposed into its several constituents.

If the object is transparent, it has to be brought between the sunlight and the prism, or between the prism and the eye, in order to determine the colors so transmitted.

Pure colors are only such as appear in a spectrum from a minimal slit; hence, the reaction of an individual for pure colors must be examined by means of the spectrum, which has either to be thrown upon a screen by means of a slit and prism, or directly upon the retina with the spectroscope, the latter method being better and far less tedious.

The points to be investigated at the examination are: whether the individual sees the *whole* spectrum, as far as a normal eye is able to perceive it, or whether it appears to him shorter at one end. Which part of the spectrum appears brightest or darkest. Whether he sees a certain

* "Chromatoptometry": The examination of the eye as to its sensibility for rays of light, of various wave-lengths.

portion of the spectrum dark (thus manifesting the entire loss of perception of a given color). He is then requested to relate the succession of the colors; or the parts of the spectrum which appear to the normal eye as the fundamental colors are isolated. This is done by pushing a black screen over the spectrum, with a slit so narrow as only to permit the color in question to be seen through it.

In all instances the situation of the color has to be indicated, in its relation to the Fraunhofer's lines.

We have only to add that the spectrum colors can be mixed for the purposes of this examination.*

Quite recently, Dr. J. Stilling† has introduced in the examination of the eye, in cases of partial color-blindness, the employment of the bright metal lines. In a number of such cases, the bright lines of Lithium, Sodium, Calcium, and Thallium, have been used by him, the individual under examination indicating the color of the line presented through the spectroscope.

We have previously mentioned colored glass as a means of examining the reaction upon colors. Colored glass, when employed for spectacles, ought to be examined with the spectroscope in reference to the colors they transmit.

Neither the apparent color-tint, the brightness of the transmitted light, nor the thickness, for instance, of a blue glass, permit the anticipation as to which particular colors will be absorbed or transmitted; the spectroscope alone can determine the matter.

* Snellen und Landolt, Ophthalmometrologie (Graefe und Saemisch, Handbuch der Augenheilkunde, III. 1, 1874). Maxwell (Philosophical Magazine, 1861). Experiments on color. 1855. Helmholtz, Physiologische Optik, 1867.

† Stilling, J., Beiträge zur Lehre von den Farbenempfindungen. Stuttgart, 1875. (Beilage zu den Klin. Monatsblättern f. Augenheilkunde.)

## CONCLUDING REMARKS.

SPECTRUM ANALYSIS must not, for a moment, be considered as a modern Philosopher's Stone, though it has already rendered considerable service in the hands of physiological chemists; and will doubtless lead us to still more valuable results.

We are especially indebted to it for the progress in our acquaintance with the coloring matters of the animal body, the complicated composition and mutability of which does not admit of ordinary chemical analysis.

It should, however, not be forgotten that the absorption phenomena give (properly speaking) nothing but a detailed description of color; and that they, therefore, can only be considered and employed as an aid to chemical analysis, their chief value consisting in indicating the direction in which the chemist must pursue his researches.

On the other hand, these phenomena will, without doubt, enable the physician to diagnose pathological alterations occurring in blood or other animal fluids, wherever the absorptive characters of the respective pigments are once definitely established. And if Spectrum Analysis be not able to solve every forensic problem, it will still be useful in toxicology and in experimental pathology, where the *causes* are *not* sought for, to demonstrate the mode of action and effect.

We have, in the preceding pages, endeavored to furnish

the student, and more especially the practical physicians, (who are already preoccupied with other and perhaps more important branches of study) with a compilation of the more salient points of medical Spectrum Analysis, without pretending a thorough exhaustion of the subject.

NEW YORK, December, 1875.

## SUPPLEMENTARY NOTE TO CHAPTER VI.

While this treatise was in press, A. Schmidt's Essay, "Ueber die Dissociation des $O_2$-Hb im lebenden Organismus," Jena, 1876, was received, where PREYER'S method of examining the blood with exclusion of air is described, by which, among other results, the HS band was observed in the blood of warm-blooded animals poisoned by HS.

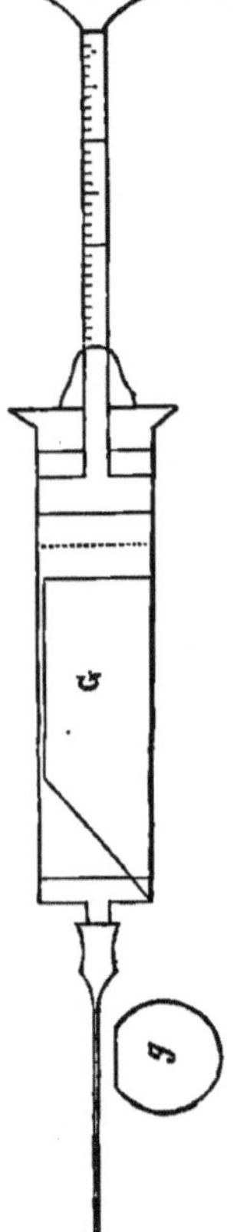

FIG. 7.

### PREYER'S APPARATUS.

A Pravaz syringe, into which a glass cylinder *G* has been fitted, is filled with rock-oil (naphtha, Steinoel), which does not contain O, and does not deprive the blood of its O for many hours. Without allowing air to enter the tube, the piston is then put into place, and the superfluous oil thrown out through the canula.

The glass cylinder must have the form indicated by *G* and *g*. It must be cut obliquely at the lower end and plane at one side.

The thin stratum of blood between the glass wall of the syringe and the plane side of the cylinder serves as object for the micro-spectroscope.

As soon as a sufficient quantity of blood from the heart of the animal has been drawn into the syringe, the point of the canula has to be closed by immersion in warm wax.

www.ingramcontent.com/pod-product-compliance
Lightning Source LLC
Chambersburg PA
CBHW022002190326
41519CB00010B/1361

* 9 7 8 3 7 4 2 8 2 1 6 2 1 *